*Understanding the Elements of the Periodic Table*™

# PLUTONIUM

Greg Roza

94    244

Pu

rosen publishing's
rosen
central®

New York

*For Allen*

Published in 2009 by The Rosen Publishing Group, Inc.
29 East 21st Street, New York, NY 10010

**Library of Congress Cataloging-in-Publication Data**

Roza, Greg.
Plutonium / Greg Roza.—1st ed.
    p. cm.—(Understanding the elements of the periodic table)
Includes bibliographical references and index.
ISBN-13: 978-1-4042-1781-2 (lib. bdg.)
1. Plutonium. 2. Periodic law—Tables. 3. Radioactivity. I. Title.
QD181.P9R69 2009
546'.434—dc22

                                                    2007047903

*Manufactured in Malaysia*

**On the cover:** Plutonium's square on the periodic table of elements. Inset: The atomic structure of plutonium.

# Contents

# Introduction

Like all elements up to fermium (Fm), plutonium (Pu) appears to be created when stars explode. Because plutonium breaks down into lighter elements over the course of millions of years, most of the natural plutonium on Earth decayed long ago. Therefore, essentially all of the plutonium on Earth has been created in laboratories during the past sixty years.

As the age of space exploration blossomed in the mid-1900s, scientists recognized plutonium's potential as a fuel for spaceships journeying into the depths of space. Over the years, it was used to power numerous manned and unmanned space explorations. In 1997, scientists from the United States and Europe launched the *Cassini-Huygens* space probe. *Cassini-Huygens* traveled 2.2 billion miles (3.5 billion kilometers) in just under seven years to reach Saturn. The *Huygens* probe separated from the *Cassini* orbiter and landed on the surface of Saturn's moon Titan. Since July 2004, *Cassini* has been exploring Saturn and its many moons from orbit.

The *Cassini* orbiter contains 72.3 pounds (32.8 kilograms) of plutonium dioxide ($PuO_2$) fuel—the most plutonium launched into space at the time. This launch was strongly protested by antinuclear activists. Many people were afraid that an accident during the launch could be disastrous. Scientists provided extensive evidence and testing data to prove that the

The *Cassini-Huygens* space probe is being prepared for its mission at the Kennedy Space Center, Florida, on August 22, 1997.

chances of an accident occurring were incredibly slim. They went ahead with the launch despite some continued objections.

*Cassini* has sent back abundant information about Saturn. Scientists planned for *Cassini* to function until 2008, but they now think it may last and be able to transmit data until 2012. This feat would not have been possible without plutonium fuel.

Plutonium plays a special role in the exploration of space. It is also used to create energy in nuclear power plants. Despite these useful abilities, plutonium is known to be a highly dangerous and controversial element because it can also be used to make nuclear weaponry. Sufficient exposure to plutonium can make people deadly sick. Despite the dangers, more than 2,000 tons (1,800 metric tons) of the element have been created over the past seventy years. Plutonium can be handled safely with the appropriate skills and equipment.

# Chapter One
## The History of Plutonium

It is difficult to discuss plutonium without first mentioning uranium (U), the element from which plutonium is made. In the eighteenth century, German silver miners discovered a black mineral they called pitchblende. Today it is called uraninite or uranium dioxide ($UO_2$). In 1789, German chemist Martin Heinrich Klaproth analyzed pitchblende and identified that it contained a new element. He named it uranium after the planet Uranus.

In 1896, quite by accident, the French physicist Henri Becquerel discovered that uranium emitted strange rays. In 1903, Becquerel and Polish-French physicist Marie Curie received the Nobel Prize for Physics for the discovery and study of this radiation. Their work was just the beginning of the exploration of radioactive chemical elements.

Shown here is a sample of pitchblende, which today is more commonly called uranium dioxide.

# The Search for Plutonium

With the discovery of uranium and radiation, scientists began to wonder if there were ways to use them to make even heavier elements. No element heavier than uranium had been discovered in nature, but many scientists thought it would be possible to make them in a laboratory. By the 1930s, many scientists began searching for the elusive "transuranic elements," or those elements that come after uranium on the periodic table. They used a new technique of bombarding atoms with nuclear particles, which had been recently discovered. It had been proven that this could turn one type of atom into a different one. Attempts to form plutonium in this way did not appear to be successful.

Some scientists thought that the particles required to synthesize heavier elements needed to be different from neutrons. Some also thought the particles did not travel fast enough to create the reactions they wanted with the technique they had been using. Scientists at the University of California at Berkeley under the direction of Ernest O. Lawrence developed a groundbreaking invention that addressed the apparent problem. This invention, called a cyclotron, accelerates charged particles such as protons by spinning them in a spiral, where they repeatedly encounter alternating voltages that continually increase their speed and energy.

In the 1940s, Ernest Lawrence examines the cyclotron he invented in a University of California laboratory.

# Plutonium Is Born

On December 14, 1940, American chemists Glenn T. Seaborg, Arthur C. Wahl, and Joseph W. Kennedy created plutonium for the first time by bombarding uranium-238 with deuterons (a proton and a neutron) in the cyclotron following the inspiration of Edwin M. McMillan. This created an isotope of an element called neptunium (Np) that had recently been discovered. The team of scientists noticed that the new element emitted radiation in the form of beta particles, which are electrons. This form of radiation had not yet been discovered. On February 23, 1941, they were able to prove that the experiment had produced chemical element 94—

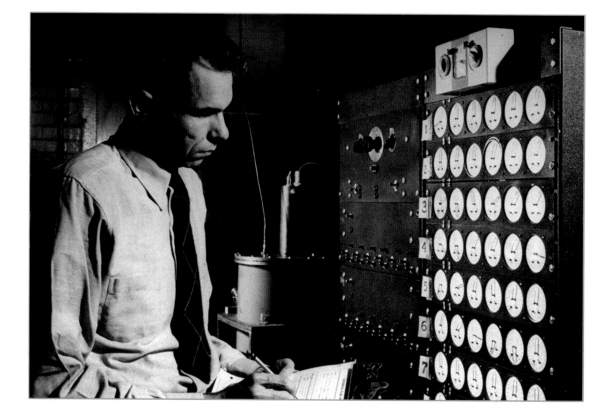

In this photo from 1946, Glenn Seaborg records meter readings on the cyclotron at the University of California. Seaborg was the primary discoverer or codiscoverer of ten chemical elements, including plutonium.

plutonium. They decided to name the element for Pluto, the dwarf planet. This followed the naming of uranium and neptunium after the planets Uranus and Neptune.

By August 1942, Burris B. Cunningham and Louis B. Werner at the University of Chicago had created enough plutonium to see with the naked eye. Soon, they had enough to weigh (3 micrograms). The room in which this happened—room 405 of the George Herbert Jones Laboratory—was declared a National Historic Landmark in 1967.

## Glenn T. Seaborg

Glenn T. Seaborg was born 1912. When Seaborg was just fourteen, he knew he wanted to be a scientist. After doing very well at the University of California at Los Angeles (UCLA), he attended the University of California at Berkeley as a graduate student of chemistry, earning his Ph.D. in 1937. He stayed at Berkeley after his doctorate, first as a lab assistant, then as an instructor of chemistry, and finally as a professor. There he was fortunate to work with some other well-known and brilliant scientists, particularly Gilbert N. Lewis, his mentor.

Over the next decade, Seaborg helped isolate and produce ten elements heavier than uranium, including plutonium. His discoveries led to a revision of the periodic table. He was honored for his scientific work in 1951, when he won the Nobel Prize for Chemistry. He was a teacher and scientist, and he eventually became the second chancellor of Berkeley. For decades, Seaborg was a strong advocate for the use of nuclear energy for peaceful purposes. Seaborg wrote hundreds of influential articles and books and won many awards. He has even had an asteroid and a chemical element, seaborgium (Sg), named after him.

# Secrecy, War, and Power

Seaborg, Wahl, and Kennedy had initially set out to publish their findings, but the U.S. government cautioned against doing so. World War II had been raging in Europe since 1939. Other countries, including Germany, Russia, and Japan, were experimenting with radioactive materials, too. They knew these materials could be used to create destructive explosions, and whoever perfected the technology first would have a great advantage in the war. As a result, the project was carried out in secrecy.

In the 1940s, American scientists began producing large quantities of plutonium as part of a program aimed at developing the first atomic bomb.

American physicist Robert Oppenheimer *(left)* points to a photo of an atomic bomb explosion, which used plutonium. Glenn Seaborg *(far right)*, who helped to create plutonium, looks on.

This program was called the Manhattan Project. Secrecy surrounded the project. Other nations were also trying to perfect an atomic bomb, but the United States was able to do it first. By 1945, the United States had several kilograms of plutonium, which was enough to build three atomic bombs. On August 9, 1945, the United States detonated an atomic bomb containing plutonium over Nagasaki, Japan.

The use of atomic weaponry brought World War II to a close, which probably saved thousands of lives on both sides. It was, however, just the beginning of an era of hostility and distrust between countries that would become known as the Cold War. The United States, Russia, and China were the first countries to begin producing plutonium for use in nuclear weapons. They continued to do so until the 1990s. At one time, the United States had eighteen plutonium production plants, all of which are now closed down. Russia and China also led ambitious plutonium programs. Today, nine countries have stockpiles of nuclear weapons. Some have recently sought to attain or make their own nuclear weapons.

# The Use of Plutonium Today

Today, plutonium is used as a fuel in nuclear reactors. Many people feel nuclear fuel is an important energy source and that it will continue to play an important role. Plutonium is also used as a fuel in the space program. It was once even used as an energy source for pacemakers.

Despite its usefulness, plutonium is also considered to be one of the most dangerous elements. Scientists estimate that Earth's atmosphere is polluted with more than 22,000 pounds (10,000 kg) of plutonium due to atomic bomb explosions and nuclear pollution. In addition, there is enough weapons-grade plutonium (which typically contains 93 percent or more of plutonium-239) in the world to wage countless wars. In the future, countries need to work together to find a peaceful and safe way to resolve conflicts without the production and use of nuclear weapons.

# Chapter Two
## Plutonium and the Periodic Table

Take a look at the objects around you: your chair, desk, bed, even the book in your hands. These things probably look very solid to you. You might be shocked to learn that everything in the universe—yes, even you—is made up of trillions upon trillions of tiny pieces called atoms. They are even in the air around you. Although they are unbelievably small, atoms are mainly made up of space. Within that space are even smaller pieces of matter called subatomic particles.

## Getting to Know Subatomic Particles

Atoms are the smallest pieces of matter that can exist by themselves and still retain the distinct properties of the element of which they are made. Atoms are made up of smaller particles; the most important are protons, neutrons, and electrons. At the center of every atom is a body of particles called the nucleus. The nucleus is made up of protons and neutrons. These are the particles that give an atom most of its weight. Protons are positively charged particles. The number of protons an atom has is its atomic number. Each atom of plutonium, for example, has 94 protons, so its atomic number is 94. The number of protons in atoms is important because this determines properties of the elements and where they are located in the periodic table.

This is a shell model of an atom of plutonium. In reality, there is far more space between the electrons and the nucleus. A plutonium atom has ninety-four electrons in seven energy shells.

Neutrons do not have an electrical charge. Although atoms of a particular element always have the same number of protons, they do not always have the same number of neutrons. Plutonium's most stable form has 150 neutrons. However, plutonium atoms are known with as many as 153 and as few as 134 neutrons. These different atoms are still considered plutonium atoms, and they all occupy the same place on the periodic table. The atoms of a specific element with different numbers of neutrons are called isotopes. Scientists have identified twenty plutonium isotopes.

Electrons orbit the nucleus in overlapping layers called shells. Usually, an atom has the same number of electrons as protons. Electrons have almost no weight at all and they have a negative electrical charge. This charge is attracted to the positive charge of the protons to help keep them in orbit around the nucleus. Sometimes, however, electrons jump from one atom to another. An atom that loses one or more electrons has an overall positive charge because of a greater number of positively charged protons than negatively charged electrons. An atom that gains one or more electrons has an overall negative charge because of the excess electrons compared to protons. An atom with either electrical charge is called an ion.

# The Periodic Table

Scientists have long tried to formulate a system for organizing and studying the chemical elements. The periodic table that we are familiar with

The portrait here depicts Dmitry Mendeleyev hard at work in his laboratory in St. Petersburg, Russia.

today was formulated over many years by several scientists, including the Russian chemist Dmitry Mendeleyev in the 1860s. Mendeleyev wrote important information about each known element on separate cards. He assembled these cards in order of their atomic weights. Once he did this, he noticed that certain chemical and/or physical characteristics recurred in regular intervals, or periods. Mendeleyev used this ordering to arrange the elements in a table, even leaving gaps where undiscovered elements might fit in the future. Over the years, many of his predictions of the properties of those undiscovered elements were proven to be true as newly discovered and synthesized elements filled the gaps he had left. The periodic table as it is now known evolved from this early one.

## Periods, Groups, and Series

The periodic table groups elements in rows and columns. Rows on the periodic table are called periods. The atomic number increases numerically as you read across a period. Periods begin with an alkali metal (group 1) on the left, and end with a noble gas (group 18) on the right. All elements in a period have the same number of electron shells, although those shells contain a different number of electrons.

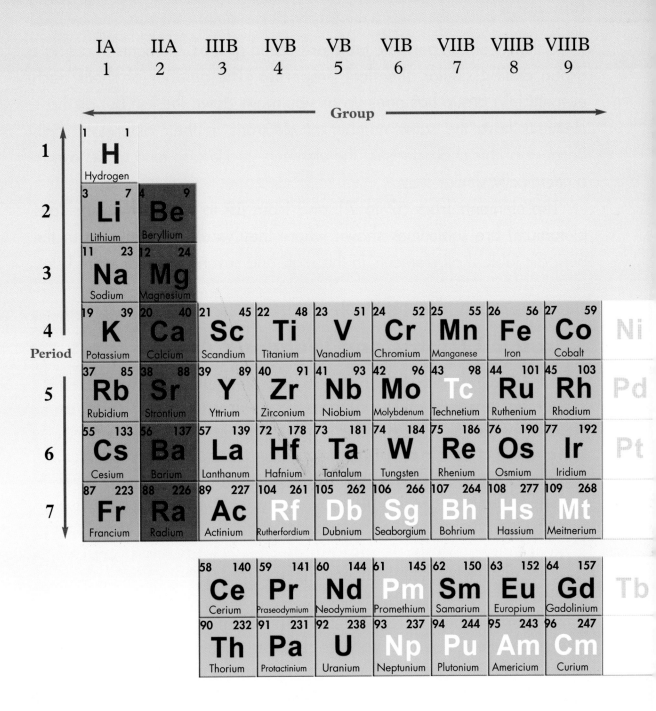

The left half of the periodic table is shown here. Plutonium appears in the bottom row (period 7), fifth element from the left, with the actinides. *(See pages 40–41 for the complete periodic table.)*

Columns on the periodic table are called groups. Elements in a single group exhibit similar chemical properties. The number of shells each element in a group has goes up as you move down the list, but all those elements have the same number of electrons in their outermost shell. Because of this characteristic, the elements in a group tend to behave in a chemically similar way.

The elements from 57 to 71 and from 89 to 103 (which includes plutonium) are sometimes shown where they would normally go in the periodic table if all elements in the sixth and seventh periods were listed

## Atomic Weights and Weighing Atoms

The periodic table displays most chemical elements with their average atomic weights (expressed as atomic mass units, or amu), which is the average mass of all isotopes of an atom weighted by their naturally occurring abundances. However, for synthetic elements like plutonium, the weight shown is that of the most stable isotope of the element, which is also frequently rounded to the nearest digit. For plutonium, the most stable isotope is plutonium-244 (94 protons and 150 neutrons), which has an atomic weight of 244.0642, often shown as 244 in parentheses.

Atomic weights are important to chemists. It is impossible for scientists to count every single atom involved in a chemical reaction. Therefore, scientists keep a careful watch on the total weight of each reactant or ingredient involved in experiments (for which there are usually many more than a trillion atoms involved in a typical reaction). Scientists use these to help determine the identity and amount of products made in the reaction. To do this, they carry out precise calculations using atomic weights.

numerically. This is called the long form of the periodic table and is usually too wide to show on one page. Therefore, more frequently all, or all but one, of these elements are placed separately below the table in two rows to make the table more compact. This arrangement also reflects that there are some significant similarities in the properties of the elements listed in these rows. The first row is often called the lanthanides and the second row the actinides. Plutonium is an actinide.

## What Are Actinides?

During their research, Seaborg and his team had difficulty isolating americium (Am) and curium (Cm). They noted that these elements did not have characteristics that scientists had assumed they would have. This led them to believe that these elements should be part of a new series of elements. Therefore, Seaborg suggested making changes to the periodic table of the time to appropriately accommodate those elements. In 1945, he published his suggestion; this is one of the accomplishments he was recognized for in his receiving of the Nobel Prize.

The actinides include: actinium (Ac), thorium (Th), protactinium (Pa), uranium, neptunium, pluto-nium, americium, curium, berkelium (Bk), californium (Cf), einsteinium (Es), fermium (Fm), mendelevium

In 1997, Glenn Seaborg holds a sculpture of the periodic table that one of his fans made for him.

# Plutonium $^{94}_{244}$ Pu Snapshot

| | |
|---|---|
| Chemical Symbol: | Pu |
| Classification: | Rare earth element; metallic; actinide series |
| Properties: | Silvery, radioactive metal; warm to the touch |
| Discovered By: | First identified by Glenn T. Seaborg, Arthur C. Wahl, and Joseph W. Kennedy on December 14, 1940 |
| Atomic Number: | 94 |
| Atomic Weight: | 244.0642 atomic mass units (amu) |
| Protons: | 94 |
| Electrons: | 94 |
| Neutrons: | 150 |
| State of Matter at 68° Fahrenheit (20° Celsius): | Solid |
| Melting Point: | 1,184°F (640°C) |
| Boiling Point: | 5,846°F (3,230°C) |
| Commonly Found: | In very small traces in Earth's crust; synthesized in nuclear reactors and by particle accelerators |

(Md), nobelium (No), and lawrencium (Lr). Some scientists continue a debate as to whether or not actinium should be a member of the actinides.

Thorium and uranium are the only actinides that occur naturally in Earth's crust in relatively large amounts. Small traces of neptunium and plutonium have been found in uranium deposits. Actinium and protactinium are sometimes found as a result of the decay of thorium and uranium isotopes. The rest of the actinide elements are all synthetic. Most of the actinides are challenging to study because they are radioactive. Many of them last for only an instant before decaying to form other elements. This makes them dangerous to handle unless you know what you are doing and have the proper equipment.

# Chapter Three
# Plutonium and Radioactivity

Radioactivity is particles and/or energy emitted, or radiated, by the nuclei (plural for nucleus) of unstable atoms. When the number of protons or neutrons in a nucleus grows too large or the ratio of the two is not right, the forces holding it together can no longer withstand other forces trying to rip it apart. The result is an unstable nucleus, which emits radiation in the form of subatomic particles and/or energy. Radioactivity can be helpful and dangerous. It can be used to generate power. It can also be used in medicine for invaluable tests or cancer treatments. However, too much exposure to radiation can make people very sick, and certain radioisotopes (radioactive isotopes) have been used to make devastating weapons.

## Break It Down

Unstable atoms have too many protons and/or neutrons packed into their nuclei. These types of atoms have unstable nuclei and thus are called radioactive. Most chemical elements have some isotopes that are radioactive. Some are synthetic and some occur naturally. Scientists call radioactive atoms of these isotopes radionuclides.

Many radioactive decay processes result in the formation of a different element from that of the original element. This is a process called

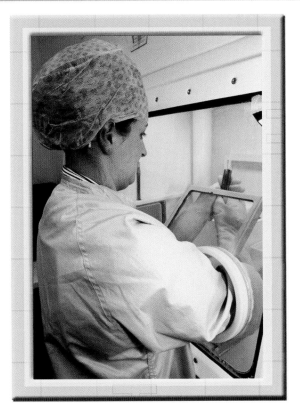

A health worker prepares a dose of radionuclides in a special sealed box, which keeps her from being contaminated with radiation.

transmutation. The new elements can be radioactive, too, such that they also decay. If this occurs for several steps, the entire process is called a decay series. Eventually, all radioactive elements decay to form stable elements.

For example, carbon-14 (C-14) is a radioisotope of carbon created in the atmosphere as a result of cosmic rays. The nucleus of an atom of carbon-14 contains six protons and eight neutrons. Carbon usually has six neutrons. The two extra neutrons in carbon-14 cause the nucleus to become unstable and it decays. When an atom of C-14 decays, it changes into nitrogen-14 (N-14), which is a stable isotope of nitrogen.

# Radioactive Decay

As radioactive elements decay, they release subatomic particles and/or bursts of energy to eventually end up in a more stable form. We call this radioactive decay. You might be most familiar with three forms of radiation called alpha particles, beta particles, and gamma rays. Most radioactive substances produce either alpha or beta particles. Nearly all radioactive substances also produce gamma rays. Great care must be taken with radioactive elements and compounds because radiation is very

This is a computer-generated image of radioactive decay. A heavy element *(upper left)* decays by releasing an alpha particle. It eventually becomes a lighter element *(lower right)*.

energetic and can damage living tissue. Some forms of radiation are more damaging than others.

## Alpha Decay

Alpha decay occurs when a nucleus releases alpha particles, which are made up of two protons and two neutrons. This is the same as the nucleus of an atom of helium (He). Alpha particles have a positive two charge.

## Beta Decay

Beta decay occurs when a nucleus emits a beta particle, which is an electron. A beta particle forms when a neutron in a heavy nucleus breaks down and forms a proton and an electron. The electron (beta particle) then escapes the atom.

## Neutrons and Deuterons

Some radioactive processes release neutrons. Neutrons are commonly used in radiation experimentation. Scientists also use a particle called a deuteron, which is the nucleus of an atom of deuterium that has a plus one charge. Deuterium is an isotope of hydrogen, which has one proton and one neutron. Deuterons were the particles used to produce the first plutonium atoms.

## Gamma Decay

When a nucleus emits an alpha or beta particle, it is usually left in an excited state, meaning that it has extra energy. It releases this energy in the form of gamma rays. This is a type of electromagnetic light, as are X-rays, ultraviolet rays, microwaves, and radiowaves. Gamma rays and X-rays are high-energy electromagnetic waves. Radio waves are low-energy waves.

Gamma rays are all around us all the time. They come from naturally radioactive elements in the earth, and from distant stars and galaxies. Gamma rays can harm living cells and lead to illnesses such as cancer.

This is an artist's depiction of a gamma ray burst. In this case, the burst results from the collision of two stars. The immense gravity between the stars emits an enormous amount of energy in the form of gamma rays.

Radioactive materials must be stored in containers that will not allow deadly gamma rays to escape. Ironically, gamma rays can also be used to destroy some types of cancer.

## Positron Emission

Protons in radioactive atoms can sometimes turn into neutrons. When this happens, the atom releases a positively charged beta particle called a positron. Positrons are identical to electrons in mass, but they have a positive charge.

# Half-Life

Radioactive decay occurs at different rates depending on the radioisotope. The nuclei of some radioisotopes last for millions of years. Others last for just a fraction of a second. An isotope's decay rate is measured by the time required for one-half of the nuclei in a given sample of the isotope to disintegrate. This quantity is known as the half-life for that radioisotope.

The half-life for a particular radioisotope is always the same. Let's say you have 100 grams of a radioactive substance with a half-life of one day. In twenty-four hours, 50 grams of the substance will have decayed, and 50 grams of the original substance will remain. In another twenty-four hours, 25 grams of the original substance will remain, which is half of what was left after the first day.

A radioactive material decays essentially forever because only half of its nuclei decay during each half-life period. This always leaves half of the original radioactive material behind. As a result, radioactive elements stay at least slightly radioactive forever. However, at some point the radioactivity is very slight and can no longer be detected. This can be a very long time for an element like plutonium-244—the most stable isotope of plutonium—which has a half-life of 82,000,000 years.

# Radioactivity and Plutonium

Plutonium is a product of the decay of uranium. Plutonium forms when a neutron is incorporated into the nucleus of a U-238 atom, which forms U-239 (this process is known as neutron capture). U-239 undergoes beta decay to form Pu-239. Other isotopes of plutonium are made by different pathways, most of which involve the decay of other uranium or plutonium isotopes.

Pu-239 has a half-life of 24,000 years. It is warm to the touch because of the energy it constantly releases due to radioactive decay. Large amounts of the element can even boil water. Plutonium breaks down

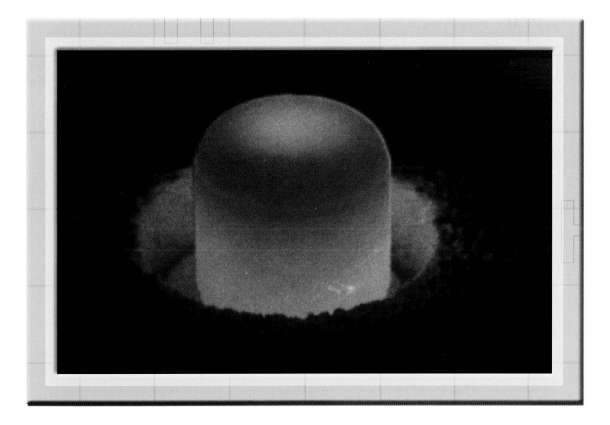

The glowing mass in this photo is a plutonium-238 sphere. Spheres such as this one are used to power satellites and space probes. This particular sphere was actually left on the moon by the Apollo space mission astronauts.

by alpha particle decay. Because alpha particles are the least penetrating form of radiation and can be shielded with many thin materials, small amounts of plutonium-239 can be handled safely with a pair of rubber gloves. The real danger with plutonium-239 radiation comes from inhaling plutonium dust because there is no barrier to stop the alpha radiation from doing internal damage.

# Chapter Four
# Working with Plutonium

**P**ure plutonium is a dense, silver metal. When exposed to oxygen or air (which contains 20 percent oxygen), it quickly develops a dull yellow layer of plutonium dioxide ($PuO_2$) tarnish. Plutonium exists in extremely small quantities in natural uranium deposits. Depending on the solid form that plutonium adopts, it can be as brittle as glass or as malleable as aluminum.

## Plutonium Allotropes

Allotropes are different forms of an element in the same state of matter. Diamond and graphite are two common allotropes of carbon. What makes these two forms different is the manner in which their

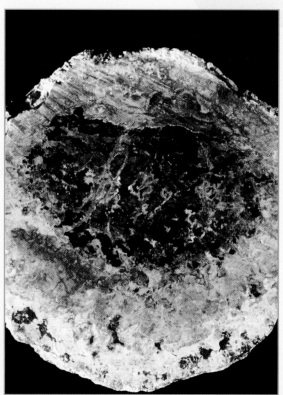

Shown here is a sample of uraninite—a mineral rich in uranium. Small traces of plutonium are sometimes found in uraninite deposits.

atoms are bonded together. The atoms in graphite are arranged in layers with weak bonds. The atoms in diamond form very regular, repeated bonds, making it the strongest natural substance on Earth.

Plutonium has six allotropes under normal circumstances and an additional allotrope under pressure. Each has a different bonding pattern. They differ mainly in their densities. Plutonium's allotropes make it unpredictable and sensitive to changes in pressure, temperature, and chemistry. When heated and cooled, plutonium can unexpectedly change from one allotrope to another, sometimes increasing or decreasing in density by as much as 25 percent. In addition, some forms are very malleable and others are quite brittle. Because of this, plutonium is challenging to work with.

# Making Plutonium

Because so little plutonium exists in nature, scientists create it in laboratories at a large expense. It is therefore considered an artificial element, rather than a natural one.

Large-scale plutonium production occurred in the 1940s, first at a pilot plant in Oak Ridge, Tennessee, and then in the first large-scale reactor at Hanford, Washington. The production of plutonium is achieved through a series of chemical reactions. Natural uranium-238 is bombarded by neutrons, which changes it into U-239. After beta decay (the half-life for U-239 is 23.5 minutes), neptunium-239 is formed. The Np-239 decays to form Pu-239, which has a half-life of 2.4 days. Other plutonium isotopes are manufactured in similar ways.

Plutonium is created in a nuclear reactor—a place where controlled nuclear reactions take place. Plutonium-generation reactors or power plants that primarily use uranium as fuel are two types of nuclear reactors. Many submarines, aircraft carriers, and ice breakers are powered by nuclear reactors. Even though no countries are purposely creating weapons-grade

The USS *Montpelier* is a nuclear-powered, fast-attack submarine. Uranium is most often used as a fuel for nuclear submarines. Plutonium is a by-product of the nuclear reactions needed to create power for the submarine.

plutonium at this time, plutonium-containing waste from nuclear power plants continues to increase every year. Scientists have yet to discover a safe and inexpensive way to dispose of plutonium waste that is guaranteed not to leak over the course of thousands of years.

## The Power of Plutonium

A specific quality and quantity of radioactive material is needed in order to use it for nuclear power or atomic weapons. The process that provides the energy is called nuclear fission. Fission is the splitting of an atom into smaller parts, which also results in the release of heat. For fission to occur, the target nucleus typically requires the absorption of a neutron. Fission is sometimes referred to as "splitting an atom."

Only some nuclei are susceptible to nuclear fission. Plutonium-239 is such a fissile radionuclide. When a neutron splits a plutonium atom into smaller pieces, heat is released, and more free neutrons are created. Under the right circumstances, the newly freed neutrons can then split other plutonium atoms. This releases more heat and more free neutrons, which continues the fission activity. Scientists call this a chain reaction. It can occur in a fraction of a second.

A chain reaction produces an immense amount of heat. When a nuclear weapon explodes, the chain reaction happens quickly and the heat released is devastating. Nuclear reactors, however, slow the chain

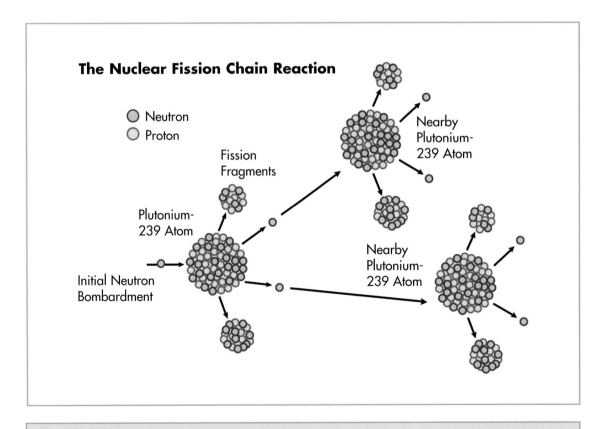

This diagram illustrates a neutron splitting a Pu-239 atom. Among the fragments created by this collision are two more neutrons, which can go on to split two more atoms of Pu-239, causing a chain reaction.

# Atom Smashers

Particle accelerators (or atom smashers) use an electric field to drive electrically charged subatomic particles to great speeds. The accelerated particles can then be aimed at a target nucleus, such as uranium, in order to achieve a nuclear reaction. X-ray machines and some televisions use low-energy particle accelerators that do not generate enough energy for nuclear reactions.

Scientists trying to create elements heavier than uranium first developed and used circular accelerators called cyclotrons. The largest circular accelerator yet is nearly finished. It is the Large Hadron Collidor located on the border between France and Switzerland. It has a circumference of 17 miles (27 km).

There are also linear accelerators that consist of an electric field around a straight-line path. The first accelerators of this type were only about 6 feet (1.8 meters) long. Today, the longest linear accelerator—the Stanford Linear Accelerator Center at Stanford University—is 2 miles (3.2 km) long. Longer accelerators can produce more energetic particles, which give scientists additional experiments that can be done with the accelerator besides the synthesis of new elements.

This is an aerial view of the Stanford Linear Accelerator Center (SLAC) in Stanford, California.

reaction down so the heat is released slowly and steadily. In order for a chain reaction to take place, the sample of plutonium must be large enough to sustain it. A sample that is at or over that size is said to be at critical mass. The critical mass of a sample of plutonium depends on several criteria, including purity, size, density, and shape, as well as the material surrounding it. In addition, some isotopes release more neutrons than others, making them able to reach a critical mass with less material. Pu-239 reaches critical mass more easily than other isotopes.

# Plutonium Compounds

There are several notable plutonium compounds and alloys, but they are all used in the same manner in which plutonium metal is used. Some can be used in nuclear reactors and nuclear weapons, while others are by-products of nuclear reactions. Plutonium dioxide can be used as reactor fuel when mixed with uranium dioxide ($UO_2$). Plutonium carbides (plutonium and carbon compounds) can be used in breeder reactors. Plutonium fluorides (plutonium and fluorine (F) compounds) are used in the production of pure plutonium metal. When plutonium is extracted from used reactor fuel, it may take the form of plutonium nitrates, which are compounds containing plutonium, nitrogen, and oxygen.

Plutonium can also take the form of an alloy—a mixture of two or more metals, or a metal and a nonmetal. Scientists discovered that mixing plutonium with small amounts of metals like aluminum (Al) or gallium (Ga) keeps the plutonium in an allotropic form that is very malleable and can easily be worked into different shapes.

Plutonium can also appear in compounds known as mixed oxides, or MOX fuel. These are a combination of plutonium oxides with different plutonium and oxygen ratios, natural uranium, and waste uranium left over after it is used in reactors. MOX fuel is a useful way to use weapons-grade plutonium instead of disposing of it as nuclear waste.

# Chapter Five
# Plutonium in Our World

In his book *Plutonium: A History of the World's Most Dangerous Element*, physicist Jeremy Bernstein estimates that there is about 1,918 tons (1,740 metric tons) of nonmilitary plutonium in the world today. He adds that this amount increases by approximately 77 tons (70 metric tons) per year due to the plutonium created in nuclear reactors.

In addition, there is about 171 tons (155 metric tons) of weapons-grade plutonium in the world, which apparently hasn't been manufactured since the 1990s (except in North Korea). This brings the overall total to about 2,089 tons (1,895 metric tons). This is quite stunning considering that seventy years ago, hardly any plutonium existed on Earth.

These are the two nuclear reactors of the San Onofre Nuclear Generating Station on the Pacific Coast of Southern California.

This sudden accumulation raises the question of what to do with all that plutonium. If plutonium is so dangerous and we no longer need nuclear weapons, why keep making it? This chapter will try to answer that question, but in the end, it is a tricky question to answer.

# Plutonium Power

Nuclear power is the controlled use of nuclear fission to create heat and electricity. The radioactive fuel in nuclear reactors is used to boil water. Steam from the boiling water powers a turbine, which creates electricity.

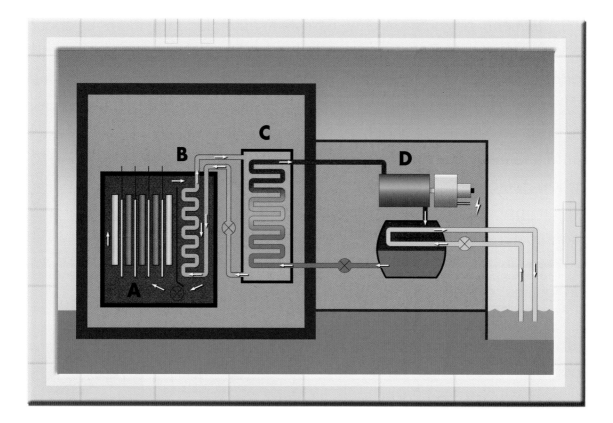

Shown here is a diagram of a type of nuclear reactor called a breeder reactor. As the fuel rods (A) release energy, they actually create more fissile material (often Pu-239). This new fissile material can be used as fuel for other nuclear reactors. Heat exchangers (B) convert water to steam (C), which powers a generator (D) and creates electricity.

Although there are several different types of nuclear reactors, this is a basic description of how most function.

Most nuclear power plants use the same fuels—U-235 (which is cheaper) or Pu-239. The fuel is submerged in water in a sealed structure called a containment building. Rods made of a material that absorb neutrons are raised and lowered into the fuel. When raised, these control rods allow the fuel to create more heat. When lowered, they absorb neutrons and decrease the number of fission reactions that occur. This reduces the amount of heat the fuel creates.

Nuclear power is a good substitute for fossil fuels. If managed properly, it is cleaner for the environment than coal, oil, or natural gas. Despite its benefits, nuclear power does have its drawbacks. One is that plutonium is a waste product of nuclear reactors.

If the nuclear fuel in a nuclear reactor is not properly cooled and the fission reaction is not controlled, the fuel can get too hot and melt. This development, called a nuclear meltdown, is economically devastating when it renders the power plant unusable. In rare cases, a meltdown can also result in an explosion and the release of nuclear fallout, causing an environmental disaster.

# Nuclear Weapons

We've already learned about fission, chain reactions, and critical mass, which are important concepts when talking about nuclear bombs. As mentioned earlier, scientists can lower plutonium's critical mass by putting it under pressure. This is how some plutonium nuclear weapons are made to explode.

Although some models differ, the following description explains plutonium's role in a nuclear bomb. At the center of the bomb is the "pit"—a sphere made mostly of Pu-239. This sphere is below critical mass. Explosive charges are placed evenly around the plutonium. It is important

# The "Fat Man"

The United States became the only country ever to use a plutonium bomb in war on August 9, 1945. On that day, the United States detonated the "Fat Man" bomb 1,650 feet (503 m) above Nagasaki, Japan, in retaliation for Japan's World War II attack on Pearl Harbor, Hawaii. This bomb used plutonium to create a massive explosion that immediately killed an estimated forty thousand people and destroyed more than two square miles of the city. Tens of thousands of more people were injured then and/or died in the years that followed.

It was estimated that the explosion was equal to 21 kilotons of TNT. As large as the explosion was, only about 20 percent of the plutonium achieved fission. It is difficult to imagine how destructive the explosion would have been had the other 80 percent achieved fission as well.

Many credit the "Fat Man" and "Little Boy" (a uranium nuclear bomb dropped on Hiroshima, Japan, three days earlier) bombs with ending World War II and bringing an uneasy peace to the world. Some even wonder if it could have saved thousands, perhaps millions, of lives had they been developed and used earlier in the war. However, the death, destruction, and contamination that resulted from the use of these nuclear bombs was immense.

The "Fat Man" bomb used plutonium and weighed about 10,200 pounds (4,630 kg).

that all of these charges explode at exactly the same time. When triggered, the charges explode inward, squeezing the plutonium into a smaller, denser ball. This pressure allows the plutonium to reach critical mass, and a chain reaction begins that creates an enormous burst of heat and energy. This reaction occurs in just milliseconds.

## Plutonium in Space

Plutonium has been used as a fuel in manned and unmanned spacecraft. These craft are powered by engines called radioisotope thermoelectric generators (RTGs). RTGs use the heat created by radioactive elements, particularly plutonium, to create electricity. You might think of them as large batteries. Most spacecraft also use solar panels to generate electricity. However, in the outer reaches of the solar system, where sunlight is much weaker, solar panels are not capable of creating enough electricity to operate the spacecraft. For this reason, RTGs are used for distant space explorations. RTGs were used during the American Apollo lunar missions, and in unmanned probes such as *Cassini-Huygens* as well as the *New Horizons* probe launched on January 19, 2006, to explore the dwarf planet Pluto.

The *New Horizons* probe blasts off into space atop an *Atlas V* rocket in 2006. The probe's generator contains about 24 pounds (11 kg) of plutonium.

# Plutonium and Health

Plutonium is actually relatively safe to handle in small amounts. However, plutonium poses a much more serious threat when ingested or inhaled. Therefore, people who work around the element wear protective gloves and suits.

Plutonium currently has no medical uses. From the early 1970s to the mid-1980s, plutonium was used to power pacemakers, but it is no longer used for this purpose. As of 2003, there were between fifty and one hundred people who still had plutonium pacemakers. When a person who has a plutonium pacemaker dies, the pacemaker is supposed to be removed and sent to the Los Alamos National Laboratory in Los Alamos, New Mexico, so the remaining plutonium can be recovered.

Due to atomic bomb tests and reactor meltdowns, and because many isotopes of plutonium have long half-lives, plutonium atoms are now always present in the atmosphere. In fact, everyone in the world probably has a few plutonium atoms in their bodies all the time. The most common form of plutonium ($PuO_2$) is nearly insoluble in water, which stops it from being metabolized by the body. If this compound is eaten, essentially all of it will be excreted from the body in the bodily wastes. However, studies have shown that when $PuO_2$ is inhaled, 20 to 60 percent remains in the lungs. A small portion of this can eventually end up in the tissues of the body. The presence of a significant amount of this alpha-particle-emitting material inside the body may then increase the risk of developing cancer, most likely in the lungs or bones (leukemia).

# Nuclear Dangers, Nuclear Safety

Although plutonium can harm people and the environment, when it is used and handled properly, plutonium fuel is much better for the environment than fossil fuels. Many countries have or are currently developing nuclear

These technicians at the Los Alamos National Laboratory in New Mexico are manufacturing plutonium pellets for use in nuclear reactors. One technician *(center)* measures plutonium oxide powder, while the other technicians press the powder into small pellets.

energy programs to replace or supplement energy systems that rely on fossil fuels. Others are suspected of developing nuclear programs for the purpose of making nuclear weapons.

In the future, the countries of the world will need to come to an agreement on how potentially dangerous substances including plutonium should be created, used, and stored to avoid further contaminating our world.

# The Periodic Table of Elements

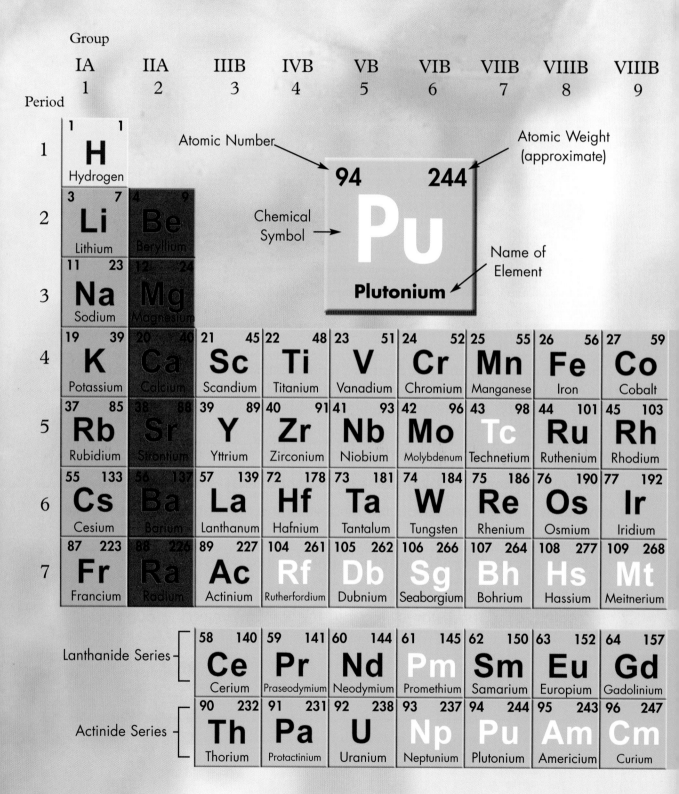

**Group**

| IA | IIA | IIIB | IVB | VB | VIB | VIIB | VIIIB | VIIIB |
|----|-----|------|-----|----|-----|------|-------|-------|
| 1 | 2 | 3 | 4 | 5 | 6 | 7 | 8 | 9 |

**Period**

Atomic Number → 94  244 ← Atomic Weight (approximate)

Chemical Symbol → **Pu**

**Plutonium** ← Name of Element

| | 1 | 1 | | | | | | | |
|---|---|---|---|---|---|---|---|---|---|
| **1** | **H** Hydrogen | | | | | | | | |

| | 3 7 | 4 9 | | | | | | | |
|---|---|---|---|---|---|---|---|---|---|
| **2** | **Li** Lithium | **Be** Beryllium | | | | | | | |

| | 11 23 | 12 24 | | | | | | | |
|---|---|---|---|---|---|---|---|---|---|
| **3** | **Na** Sodium | **Mg** Magnesium | | | | | | | |

| | 19 39 | 20 40 | 21 45 | 22 48 | 23 51 | 24 52 | 25 55 | 26 56 | 27 59 |
|---|---|---|---|---|---|---|---|---|---|
| **4** | **K** Potassium | **Ca** Calcium | **Sc** Scandium | **Ti** Titanium | **V** Vanadium | **Cr** Chromium | **Mn** Manganese | **Fe** Iron | **Co** Cobalt |

| | 37 85 | 38 88 | 39 89 | 40 91 | 41 93 | 42 96 | 43 98 | 44 101 | 45 103 |
|---|---|---|---|---|---|---|---|---|---|
| **5** | **Rb** Rubidium | **Sr** Strontium | **Y** Yttrium | **Zr** Zirconium | **Nb** Niobium | **Mo** Molybdenum | **Tc** Technetium | **Ru** Ruthenium | **Rh** Rhodium |

| | 55 133 | 56 137 | 57 139 | 72 178 | 73 181 | 74 184 | 75 186 | 76 190 | 77 192 |
|---|---|---|---|---|---|---|---|---|---|
| **6** | **Cs** Cesium | **Ba** Barium | **La** Lanthanum | **Hf** Hafnium | **Ta** Tantalum | **W** Tungsten | **Re** Rhenium | **Os** Osmium | **Ir** Iridium |

| | 87 223 | 88 226 | 89 227 | 104 261 | 105 262 | 106 266 | 107 264 | 108 277 | 109 268 |
|---|---|---|---|---|---|---|---|---|---|
| **7** | **Fr** Francium | **Ra** Radium | **Ac** Actinium | **Rf** Rutherfordium | **Db** Dubnium | **Sg** Seaborgium | **Bh** Bohrium | **Hs** Hassium | **Mt** Meitnerium |

**Lanthanide Series**

| 58 140 | 59 141 | 60 144 | 61 145 | 62 150 | 63 152 | 64 157 |
|---|---|---|---|---|---|---|
| **Ce** Cerium | **Pr** Praseodymium | **Nd** Neodymium | **Pm** Promethium | **Sm** Samarium | **Eu** Europium | **Gd** Gadolinium |

**Actinide Series**

| 90 232 | 91 231 | 92 238 | 93 237 | 94 244 | 95 243 | 96 247 |
|---|---|---|---|---|---|---|
| **Th** Thorium | **Pa** Protactinium | **U** Uranium | **Np** Neptunium | **Pu** Plutonium | **Am** Americium | **Cm** Curium |

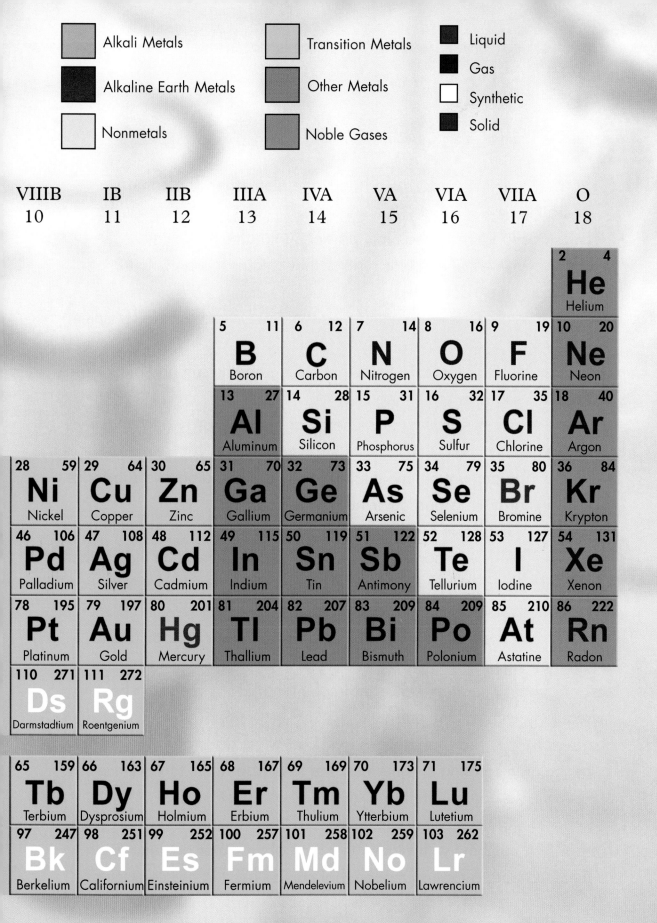

# Glossary

**allotropes** Different forms of the same element in the same state of matter.

**bombard** To launch subatomic particles into atoms.

**breeder reactor** A nuclear reactor that creates more fissile material that it originally contained.

**contaminate** To make something or some place radioactive by putting it in contact with a radioactive substance.

**cosmic rays** Subatomic particles that come from the sun and other stars.

**detonate** To make something explode.

**gamma rays** Electromagnetic energy of higher energy and frequency than X-rays.

**half-life** The amount of time it takes for half of a radioactive substance to disintegrate in a particular nuclear reaction.

**kiloton** An explosive force equal to 1,000 tons of TNT.

**leukemia** A type of bone cancer marked by an increase in the number of white blood cells in the blood. This can lead to a shortage of red blood cells, infections, and death.

**meltdown** The melting of the fuel rods in a nuclear reactor due to overheating.

**pacemaker** A battery-powered implant that keeps the heart beating regularly with electric pulses.

**radioactivity** The spontaneous emission of particles and gamma rays by the disintegration of a nucleus. The term also refers to the emissions.

**stockpile** To accumulate large quantities of something.

**synthetic** Artificial or human made.

**turbine** A machine with blades that turn when water or steam push them, causing a turning motion that can be used to create electricity.

Arms Control Association
1313 L Street NW, Suite 130
Washington, DC 20005
(202) 463-8270
Web site: http://www.armscontrol.org
A global association concerned with the public understanding of and
   support for arms control policies.

Canadian Nuclear Association (CNA)
130 Albert Street, Suite 1610
Ottawa, ON K1P 5G4
Canada
(613) 237-4262
Web site: http://www.cna.ca
A Canadian organization that promotes the development and use of
   nuclear technology for peaceful purposes.

Institute for Energy and Environmental Research (IEER)
6935 Laurel Avenue, Suite 201
Takoma Park, MD 20912
(301) 270-5500
Web site: http://www.ieer.org/index.html
The IEER offers scientific information regarding energy and the environment, written and presented with the average person in mind for the
   purpose of spreading scientific knowledge.

International Atomic Emergency Agency (IAEA)

P.O. Box 100

Wagramer Strasse 5, A-1400

Vienna, Austria

Web site: http://www.iaea.org

Originally named Atoms for Peace, the IAEA is interested in "safe, secure and peaceful nuclear technologies" for countries all over the world.

World Nuclear Association (WNA)

22a St. James's Square

London SW1Y 4JH

United Kingdom

Web site: http://www.world-nuclear.org

A global association that strives to promote nuclear power as a viable energy source for future generations.

# Web Sites

Due to the changing nature of Internet links, Rosen Publishing has developed an online list of Web sites related to the subject of this book. This site is updated regularly. Please us this link to access the list:

http://www.rosenlinks.com/uept/plut

Burgan, Michael, Nancy Cohen, Stephen Currie, and Vanessa Elder.
*Nuclear Energy* (Discovery Channel School Science). Strongville, OH:
Gareth Stevens Publishing, 2002.

Gonick, Larry, and Craig Criddle. *The Cartoon Guide to Chemistry*.
New York, NY: HarperCollins, 2005.

Jackson, Tom. *Radioactive Elements*. Tarrytown, NY: Marshall Cavendish
Benchmark, 2006.

Knapp, Brian. *Uranium and Other Radioactive Elements*. Danbury, CT:
Grolier, 1996.

Mayell, Mark. *Nuclear Accidents*. Farmington Hills, MI: Lucent
Books, 2003.

Miller, Ron. *The Elements: What You Really Want to Know*. Minneapolis,
MN: Twenty-First Century Books, 2006.

Newmark, Ann, and Laura Buller. *Chemistry*. New York, NY: DK
Children, 2005.

Oxlade, Chris. *Elements and Compounds* (Chemicals in Action).
Chicago, IL: Heinemann Library, 2002.

Stwertka, Albert. *A Guide to the Elements*. New York, NY: Oxford
University Press, 2002.

Tweed, Matt. *Essential Elements: Atoms, Quarks, and the Periodic Table*.
New York, NY: Walker, 2003.

Zannos, Susan. *Dmitri Mendeleyev and the Periodic Table* (Uncharted,
Unexplored, and Unexplained). Hockessin, DE: Mitchell Lane
Publishers, 2004.

# Bibliography

Bernstein, Jeremy. *Plutonium: A History of the World's Most Dangerous Element*. Washington, DC: Joseph Henry Press, 2007.

Greenwood, N. N., and A. Earnshaw. *Chemistry of the Elements*. Oxford, England: Butterworth-Heinemann, 2001.

Hecker, Siegfried S. "Plutonium and Its Alloys: From Atoms to Microstructure." *Los Alamos Science*. Retrieved October 26, 2007 (http://www.fas.org/sgp/othergov/doe/lanl/pubs/00818035.pdf).

Institute for Energy and Environmental Research. "Physical, Nuclear, and Chemical, Properties of Plutonium." Retrieved October 26, 2007 (http://www.ieer.org/fctsheet/pu-props.html).

Jackson, Tom. *Radioactive Elements*. Tarrytown, NY: Marshall Cavendish Benchmark, 2006.

Knapp, Brian. *Uranium and Other Radioactive Elements*. Danbury, CT: Grolier, 1996.

Los Alamos Technical Reports and Publications. "Plutonium in Use: From Single Atoms to Multiton Amounts." *Los Alamos Science*. Retrieved October 26, 2007 (http://fas.org/sgp/othergov/doe/lanl/pubs/00818005.pdf).

Piazza, Enrico. "*Cassini-Huygens*: Mission to Saturn and Titan." Jet Propulsion Laboratory. October 10, 2007. Retrieved October 25, 2007 (http://saturn.jpl.nasa.gov/overview/index.cfm).

Space Views: *Cassini*: Students for the Exploration and Development of Space. "The RTG Debate." Retrieved October 17, 2007 (http://seds.org/spaceviews/cassini/rtg.html).

U.S. Department of Energy. "The Atomic Bombing of Nagasaki." *The Manhattan Project: An Interactive History*. Retrieved October 26, 2007 (http://www.cfo.doe.gov/me70/manhattan/nagasaki.htm).

# Index

## About the Author

Greg Roza has written and edited educational materials for children and young adults for the past eight years. Roza has long had an interest in scientific topics, including chemistry, and spends much of his spare time tinkering with machines around the house. He lives in Hamburg, New York, with his wife, Abigail, and three children, Autumn, Lincoln, and Daisy.

## Photo Credits

Cover, p. 1 Tahara Anderson; p. 5 © AFP/Getty Images; p. 6 © Astrid & Hanna-Frieder Michler/Photo Researchers; p. 7 © Science Source/Photo Researchers; pp. 8, 14 © Time & Life Pictures/Getty Images; pp. 10, 33 © Getty Images; p. 17 © AP Photos; p. 21 Josh Sher/Photo Researchers; p. 22 Prof. K. Seddon & Dr. T. Evans, QUB/Photo Researchers; p. 23 © Claus Lunau/Photo Researchers; p. 25 © Photo Researchers; p. 27 © Jacana/Photo Researchers; p. 29 U.S. Navy/Department of Defense; p. 31 © Stanford Linear Accelerator/Photo Researchers; p. 34 © SPL/Photo Researchers; p. 36 © Corbis: p. 37 © Kin Man Hui/San Antonia Express/Zuma/Corbis; p. 39 © U.S. Dept. of Energy/Photo Researchers.

**Designer:** Tahara Anderson; **Editor:** Kathy Kuhtz Campbell
**Photo Researcher:** Marty Levick